The Williams and Kid's Sciences

Nomzamo Ngwabe
Author

Content Contributor
Sia Michelle

Illustrator
Aklima Monnaf

This is a work of creative nonfiction. Some parts have been fictionalized in varying degrees, for various purposes.

Copyright © 2022 by **Nomzamo Ngwabe**

All rights reserved. No part of this book may be reproduced or used in any form it can be electronic or chemical, this book can not be reproduced through voice recording, photocopying, creating other books, videos, etc. In a case where the content of this book is used, the publisher should be acknowledged.

Book illustrated by Aklima Monnaf

Biological Sciences

Mr and Mrs Williams have three children, Lisa, Missy, and Khanya.

Mrs Williams called her husband on a very faithful Monday morning and they talked.

Mrs Williams: husband and I are very pleased with our children; all three of them enjoyed science, with the exception of Lisa, who enjoys biological science.

Mr Williams: Yes, my dear, I believe you are correct; we must take action.

The next morning

Mr Williams: Missy, Khanya, and Lisa, where are you all breakfast is ready!

Missy: Sir!

Lisa: Sir!

Missy: Hmm pancakes, my favourites

Mr Williams: Yes i am cooking today to surprise my daughters

Lisa: Huh?

Mrs Williams: I was also surprised

Mr Williams: where is Khanya?

Lisa: I haven't seen her this morning.

Missy: I think she is still in bed.

Mr Williams: Call her for me

Immediately, Lisa hurried upstairs to wake Khanya up.

Lisa: Khanya, get up! Dad is summoning us downstairs and I think he is making breakfast

Khanya ran out of bed

Khanya: For real?
Lisa: yeah you had to see for yourself.

Both of them hurried downstairs.

Mr Williams: Khanya why were you not here when I called the others?

Mr Williams: and Lisa what took you so long?

Khanya: (stammering) I....... sorry father

Lisa: Khanya took so long that I was late in calling her down here.

Khanya: Liar!

They began arguing in the living room, and their father had to reprimand them, but their mother told him to leave them alone.

Mr Williams: What's the matter with you kids, can't you see that you're in the presence of your parents and acting like this?

Is this how you act in front of your teacher at school?

They sat down...

Mr Williams: The main reason I requested you here was to inform you about your education. We discovered that you all love science, so we decided to send you all to the same school.

Mrs Williams: due to personal reasons, we are unable to send you all to different schools; however, what do you think about it?

Lisa: But what happened to our old school, which we adore?
Khanya: yea why?

Mr Williams: I am sorry for this but you have to go to the same school

The three girls went to their bedroom.

Lisa: Why would Dad want to change our school? That does not sit well with me.

Khanya: Yeah, why would he do that?

Missy: I am okay with everything if you ask me

Both parents decided to send their three children to the same school.

The three children, Lisa, Missy, and Khanya, are all unique individuals.

Missy enjoys science, and Khanya enjoys science as well. Lisa, on the other hand, is fascinated by biological science, particularly human cells, brains, and skulls.

On a regular Monday, the three children leave for school, debating what their teacher assigned them as homework, "earth studies."

Lisa: Do you think many students will bring their projects to class today?

Missy: Who knows? Khanya Have you completed the project assigned to us by the teacher?

Khanya: All I did was conduct some initial research.

On their way to school they started discussing about earth sciences

Khanya Do you know that the SARS-CoV-2 virus causes Coronavirus Disease (COVID-19), an infectious virus.

Lisa mentioned biodiversity after Khanya finished her discussion.

Missy: Wow, what a horrible virus to be sick with.

Lisa: Biodiversity (from "biological variety") refers to the diversity of life on Earth at all levels, from genes to ecosystems, and can encompass evolutionary, ecological, and cultural processes that maintain life, and also 'bio means the living organisms, ecosystem according to my study last night.

As they were walking, they noticed a large amount of smoke coming from the woods, which reminded Missy of pollution.

Missy: Wow! that's a lot of smoke. Did you know that pollution is defined as the introduction of pollutants into the natural environment that causes negative change? Pollution can take the form of any material (solid, liquid, or gas) or energy (such as radioactivity, heat, sound, or light).

After the children had finished their discussion, they noticed an asteroid that had fallen from space glowing in the bush, and a chemistry student named Missy quickly took it.

Lisa: What exactly is that shining item in the bush?

Khanya: I think it's a ball

Missy moves closer to the glowing thing and it was an asteroid from space.

Missy: I can't believe my eyes, it's an asteroid from space!

Lisa: Let me see!

Lisa moves closer to missy and she held the asteroid in her hands

Lisa: It's so real and heavy

Khanya: I believe we should bring it to school and present it to our teacher.

When the kids saw the asteroid, they ran to school and waited until they were finished with the assembly ground.

And, when they were finished, the kids presented the asteroid in class, and the rest of the class is ecstatic to be experiencing something new.

Teacher: What a surprise! The Williams family brought an asteroid to class! Give them a round of applause, students!

CLASS: (CLAPPING)

When the class saw this, they were amazed. There were also two kids in class, John, and Smith, who brought in an alien eggs as their experiment and presented it to the class.

Jhon: We also found something.

When you gave us the homework yesterday we found an eggs in our neighborhood it was green and long.

Smith: Here is the eggs Ma'am!

The teacher saw the egg and was very surprised with the students, she took their experiments to the lab and stored them safely.

The school bell rang at 4 p.m., Missy, Khanya, Lisa went home to begin discussing issues concerning other planets.